2nd International Conference on Microbiome Engineering (ICME 19)

Boston, Massachusetts, USA
2 – 4 December 2019

ISBN: 978-1-7138-0572-4

Printed from e-media with permission by:

Curran Associates, Inc.
57 Morehouse Lane
Red Hook, NY 12571

Some format issues inherent in the e-media version may also appear in this print version.

Copyright© (2019) by AIChE
All rights reserved.

Printed with permission by Curran Associates, Inc. (2020)

For permission requests, please contact AIChE
at the address below.

AIChE
120 Wall Street, FL 23
New York, NY 10005-4020

Phone: (800) 242-4363
Fax: (203) 775-5177

www.aiche.org

Additional copies of this publication are available from:

Curran Associates, Inc.
57 Morehouse Lane
Red Hook, NY 12571 USA
Phone: 845-758-0400
Fax: 845-758-2633
Email: curran@proceedings.com
Web: www.proceedings.com

TABLE OF CONTENTS

COMPUTATIONAL MODELING TO PREDICT AND UNDERSTAND MICROBIOMES

EMERGING TECHNOLOGIES TO MANIPULATE AND STUDY MICROBIOMES

ENGINEERING OF ORGANISMS AND COMMUNITIES FOR THE ENVIRONMENT AND BIOTRANSFORMATIONS

ENGINEERING THERAPEUTIC MICROBES FOR CLINICAL APPLICATIONS

POSTER SESSION

Computational Design of Synthetic Microbial Communities Incorporating Competitive and Cooperative Interactions

International Conference
MICROBIOME
ENGINEERING

Authors:

Karkaria, B. D., University College London

Fedorec, A. J. H., University College London

Barnes, C. P., University College London

The human microbiota consists of a large number of interacting species, which enables stable coexistence and community function. Manipulating microbial composition, and the resultant metabolic output, offers a novel solution to treating inflammatory diseases such as Crohn's disease and ulcerative colitis. As a step towards this, we aim to engineer synthetic communities *in vitro,* allowing us to understand the fundamental components required to form stable microbial communities.

The existence and maintenance of microbial consortia is dependent upon interactions between the community members. Here, we demonstrate a modelling framework for simulating microbial consortia existing in a chemostat environment, interacting via quorum sensing and microcin expression, combined with crossfeeding of essential nutrients. Using this framework, we have been able to predict system configurations that give the highest probability for stable steady state community establishment, and infer the necessary design principles. This model informed approach allows us to focus our experimental efforts on the most probable candidate systems, with the potential to drastically reduce the time and resources required to build stable microbial communities.

Constraint-Based Dynamic Simulation of Complex Microbiomes with µbialSim

Authors:

Centler, F., UFZ - Helmholtz Centre for Environmental Research, Popp, D., UFZ - Helmholtz Centre for Environmental Research

The cohabitation in high-diversity consortia is the norm for microbial species in most natural and medical settings. Individual speciesâ€˜ metabolic activities both depend on and modulate their immediate environment, making the full consortium a complex system which continuously responds to its embedding environment in a self-organizing manner. For many applications, a means for targeted manipulation of these systems would be highly desirable. However, the many layers, and the high diversity of interactions in these systems make them difficult to analyze, understand, and ultimatly to control. Focussing on the exchange of metabolites between species, we present the numerical simulator code *Âµ*bialSim (pronounced 'microbialsim'), which allows for the simulation of high diversity consortia following the dynamic Flux-Balance-Analysis approach which relies on genome-scale metabolic network models of individual species. We demonstrate its capabilities by simulating a 773-species human gut microbiome. Batch and chemostat conditions can be considered and simulated trajectories provide quantitative predictions on the dynamics of compound concentrations, species abundancies, and intracellular metabolic flux distributions, enabling an in-depth analysis of metabolic exchange patterns. Once a microbial consortium of interest is implemented in *µ*bialSim, scenario simulations allow for the fast and efficient screening of intervention strategies able to nudge the consortium to desired state, for example increased methane production in anerobic digestion microbiomes, or healthy states in gut microbiomes. The faithful representation of the complexity of metabolic interactions in microbial consortia in computational models paves the way to both explore the potential and limits of control strategies, and the *de novo* design of synthetic consortia with defined properties.

Microbiome Exploration and Discovery

International Conference
MICROBIOME
ENGINEERING

Authors:

Mouncey, N., DOE Joint Genome Institute

Kyrpides, N., DOE Joint Genome Institute

Eloe-Fadrosh, E., DOE Joint Genome Institute

Roux, S., DOE Joint Genome Institute

Ivanova, N., DOE Joint Genome Institute

Wood-Charlson, E., DOE Systems Biology Knowledgebase

The cross-cutting nature of microbiome research in environmental sciences, health, agriculture, energy, and natural and built environments requires the development of new solutions and community coordination to tackle grand challenges that will accelerate basic discovery and lead to transformative advances. The exponential growth of microbiome data over the past few decades has ushered in a new era of biology, shifting the focus from descriptive observations and small-scale experimental paradigms to data-driven exploration and hypothesis generation, enabled by and relying on a rapidly growing suite of transformative data science strategies. The DOE JGI is a leader in microbiome data science through its user-centric scientific programs that have expanded the microbial tree-of-life, discovered numerous new viruses and host associations, explored microbe-plant interactions, and discovered new biochemical pathways that play important roles in nutrient cycling and microbial survival. I will discuss some of our recent scientific studies, as well as provide an overview of the new National Microbiome Data Collaborative.

Predicting Function from Composition: A Synthetic Ecology Approach to Understand Butyrate Production By Human Gut Microbiome Communities

International Conference MICROBIOME ENGINEERING

Authors:

Clark, R. L., University of Wisconsin – Madison, Venturelli, O., University of Wisconsin - Madison

Microbial community functions arise from complex systems-level behavior based on microbial interactions and environmental context. Because of the complexity of these systems, model-guided, high-throughput methods are necessary to design optimal experiments to probe system behavior and to develop solutions to problems in the realms of health (i.e. modulating microbiome disease impacts), energy (i.e. efficient and robust bioprocesses), and environment (i.e. bioremediations that restore the native ecosystem). In this work, we used synthetic ecology methods to develop a model-guided experimental approach to predict how microbial community composition impacts function, in this case butyrate production by the human gut microbiome. We first characterized growth and metabolite production of 25 genotyped human gut bacterial isolates, 5 of whose genomes encode metabolic pathways required for butyrate production. We then characterized the impact of pairwise interactions from these 25 species on growth and butyrate production of each of the 5 butyrate-producing bacteria. Using this initial dataset, we built a minimally-parameterized predictive model using tools from dynamic ecological modeling and statistical modeling to predict species abundance and butyrate production by higher-order communities. For experimental validation, we used this model to design 165 communities of 3-5 species that spanned the entire range of predicted butyrate concentrations and included every combination of the 5 butyrate producing bacteria. The model was able to accurately predict the species abundance and butyrate production of these communities in most cases. The structure of this model identifies the pairwise interactions that have the largest impact on butyrate production in these synthetic communities and thus can be used to identify general principles of butyrate production within this synthetic ecology model. The knowledge and methodologies generated in this work will be useful for designing interventions to treat patients with butyrate-deficient gut microbiomes as well as for designing microbial communities with specified functional outputs in general.

Systems Modelling of the Skin Microbiome

Authors:

Kumar, R., SASTRA Deemed University

Raman, K., Indian Institute of Technology Madras

The skin microbiome is a complex ecosystem comprising diverse microorganisms inhabit distinct microenvironments. The nature of interactions within the microbial community and with the host, have far-reaching implications for skin health and disease. In this study, we present several insights into these interactions, through a systems-level graph-theoretic analysis of the skin microbiome, by studying the metabolic networks of constituent organisms. We illustrate several dependencies amongst the skin species by considering the possible metabolic exchanges that can take place, and also look into the preferred mechanisms, if any, adopted by species to support each other. Using MetQuest, a graph-theoretic algorithm previously developed in our laboratory, we quantify the extent and nature of support one organisms offers to another within a community, which forms the basis of our community. Interestingly, we observe a clear dependence of *Staphylococci* and other species on *Corynebacteria*. We further show that the metabolic exchanges between organisms are predominated by enzyme classes such as transferases and translocases, across many genera. Our approach presents several testable hypotheses for experimental verification and is a generic approach to unravel the complexity of diverse microbiomes, as seen through the lens of metabolism.

Engineering Microbiomes Using Physical-Organic Chemistry

International Conference
MICROBIOME
ENGINEERING

Authors:

Verma, M.

The gut microbiome is complex: billions of bacteria from hundreds of species produce thousands of metabolites. How do we engineer such complex communities? Physical-organic chemistry-i.e. "the science of establishing links between the molecular *structure* and *function* of a compound in a physical system"provides one simple approach. I will share two vignettes of this physical-organic approach: the first one studies the biological activity of a short-chain fatty acid (butyrate) and the second explores the impact of carbohydrates on the microbiome. Two of the biological activities of butyrate in the colon (suppression of proliferation of colonic epithelial stem cells and inflammation) correlate with the inhibition of histone deacetylases. Structure-activity relationships were compatible with the hypothesis that butyrate acts by binding to the Zn^{2+} in the catalytic site of histone deacetylases. In the second example, carbohydrates provide a method for tuning the composition of the microbiome and building future synbiotics (probiotics + prebiotics). These studies offer a starting point for guiding the desired outcomes from a 'healthy' gut microbiome.

High-Throughput Isolation and Automated Sorting of Diverse Microbiomes Reduce Biases of Traditional Cultivation Strategies

Authors:

Watterson, B., University of Chicago, Tanyeri, M., Duquesne University, Watson, A., University of Chicago, Cham, C., University of Chicago, Shan, Y., University of Chicago, Chang, E., University of Chicago, Eren, M., University of Chicago, Tay, S., UNIVERSITY OF CHICAGO

Metagenomics has revolutionized our knowledge of the structure of gut-associated microbial communities and revealed interesting associations between health and disease. However, solely determining the taxonomic structure and functional potential through metagenomics is insufficient for understanding the mechanisms behind microbial community functioning. In order to obtain a mechanistic understanding of gut-associated health and disease, metagenomics must be supplemented with isolating pure strains from patient samples. However, traditional cultivation strategies are laborious; low-throughput; and feature biased sampling of taxa due to species being rare, slow-growing, or having specific metabolic requirements. To address these limitations, we developed a high-throughput anaerobic cultivation technology based on droplet microfluidics which stochastically encapsulates individual bacteria in droplets, cultivates the isolated organisms, and automatically sorts the droplets based on colony-density. Anaerobic droplet cultivation enabled several advantages over culture plates including: (1) an increase in the cultivated community richness by up to 410%, (2) a cultivated community representation which more closely resembles the input sample, (3) an enhancement in the number of strains cultivated which were <1% abundant in raw stool, and (4) cultivation of strains within the clinically important genus *Bacteroides*, which could not be grown on agar plates. Automated sorting of droplets for slow-growing colonies further enhanced the cultivated relative abundance of rare stool strains. In total, our method improves the cultivation of organisms broadly across phyla, including organisms which are rare and/or slow-growing, and can be used in conjunction with traditional microbiology techniques to increase the speed and efficiency of recovering species from complex microbial communities.

Remote Control of Cellular Localization in the Gastrointestinal Tract Using Magnetic Particles

International Conference MICROBIOME ENGINEERING

Authors:

Buss, M. T., California Institute of Technology, Ramesh, P., California Institute of Technology, English, M. A., MIT, Lee-Gosselin, A., California Institute of Technology, Shapiro, M. G., California Institute of Technology

Engineered probiotics have the potential to diagnose and treat many diseases linked to the gastrointestinal (GI) tract; however, their therapeutic potential is fundamentally constrained by their ability to colonize desired GI niches and specifically respond to endogenous and extrinsic stimuli. As such, a method to non-invasively control the persistence and localization of engineered probiotics can greatly improve therapeutic outcomes for GI linked disorders, in addition to being an important basic-research tool for studying gut microbial communities. Magnetic fields are uniquely suited to address this challenge because of their ability to freely penetrate tissue, however the rapid decay of fields from a magnetic source limits their broader utility for targeting cells within the GI-tract. Here we demonstrate a new approach that uses micromagnets, in concert with an external magnetic-field, in order to enhance the retention of orally administered probiotics in the murine GI tract. We show that this approach enhances the colonization of probiotic *E. coli* Nissle in mice that were not treated with antibiotics. Our cellular localization assisted by magnetic particles (CLAMP) strategy can potentially be generalized to any cell (bacterial, mammalian) or drug-containing nanoparticle of interest, and can also be combined with existing non-invasive imaging modalities, thereby facilitating remote control and sensing of engineered cells at sites of interest *in vivo*.

Ultra-High Throughput Screening of Synthetic Microbial Communities in Agriculture and Human Health

Authors:

Kehe, J., Massachusetts Institute of Technology

Ackerman, C. M., Broad Institute

Cervantes, B., Massachusetts Institute of Technology

Kulesa, A., Broad Institute of Harvard and MIT

Blainey, P. C., Massachusetts Institute of Technology

To harness the power of beneficial microbial communities in biotechnology, agriculture, and medicine, a deeper understanding of the assembly rules governing microbial communities is needed. Empirical screening of synthetic microbial consortia can systematically explore interspecies interactions and environmental dependencies; however, assembling many combinations of microbes for such screening is logistically complex and difficult to achieve on a timescale commensurate with microbial growth. The kChip is a droplets-based platform that performs rapid, massively parallel, bottom-up construction and screening of synthetic microbial communities. Here, we present a screen of ~100,000 multispecies communities comprising up to 19 soil isolates in which we identified consortia that promote the growth of the model plant symbiont *Herbaspirillum frisingense* in a manner robust to carbon source variation and the presence of additional species. Furthermore, we explore the use of kChip for screening microbial consortia constructed from human microbiome isolates.

Validation of CRISPR-Based Tools for Engineering Bacteroides Species As Living Therapeutics

Authors:

Oromí-Bosch, A., Caribou Biosciences, Inc., Berry, J. D., Caribou Biosciences, Inc., Smith, S., Caribou Biosciences, Inc., Huang, J., Caribou Biosciences, Inc., Shaw, M., Caribou Biosciences, Inc., Beljouw, S. V., Caribou Biosciences, Inc., Kanner, S., Caribou Biosciences, Inc., Kotula, J. W., Caribou Biosciences, Inc.

The gut microbiota impact human health and disease progression by interacting with the immune system, thereby modulating immune activity and efficacy of immunotherapies. Recent research has focused on identifying gut bacteria that are associated with specific disease outcomes. These organisms can be engineered to express beneficial metabolic pathways, thereby generating living therapeutics. However, current tools do not enable effective genetic manipulation of most gut microbes. CRISPR-based tools are powerful systems that may be used in diverse bacterial species, enabling genome editing and gene expression control.

We are leveraging our expertise with CRISPR-based genome engineering to develop tools that enable rapid and efficient editing of diverse gut microbes into controllable living therapeutics. We are validating these tools in *Bacteroides* species since they are abundant within the human colon and have been shown to play an important role in immune development. CRISPR-based tools may overcome the major limitations of existing genome editing methods for these microbes.

We present proof-of-concept data demonstrating deployment of Cas9 in *Bacteroides* species, which enables genomic editing of these bacteria, significantly decreasing the time and resources required compared to established methods. Catalytically dead Cas9 (dCas9) binds tightly to targeted sites in the genome, enabling gene expression modulation and control over metabolic pathways. Here, we show that dCas9 modulates the metabolism of *Bacteroides* species that, when delivered to the gut, have the potential to induce an immunomodulatory effect. Finally, we have designed a replicative plasmid system that is stable and replicates across *Bacteroides* species from human isolates. This tool facilitates heterologous gene overexpression throughout this genus.

Overall, these data highlight CRISPR-based tools that offer a unique and robust approach enabling engineering of key functional pathways in a suitable bacterial species based on microbiome-immune interaction studies. CRISPR-based editing is an essential platform for the rapid development of engineered microbial therapeutics.

A Guild-Based Metabolic Model Improves Understanding of a Medium-Chain Fatty Acid Producing Microbiome

International Conference
MICROBIOME
ENGINEERING

Authors:

Scarborough, M. J., University of Vermont

Donohue, T. J., University of Wisconsin-Madison

Noguera, D., University of Wisconsin-Madison

Microbiomes can be harnessed to produce beneficial chemicals from renewable resources. Recently, microbiome fermentations have been studied for producing medium-chain fatty acids (MCFA). Predictive and diagnostic models of MCFA-producing microbiomes, however, are limited. To address this, we constructed two metabolic models of a MCFA-producing microbiome fed residues from lignocellulosic biorefining. The first model is a single-unit model (iFerment156) that contains a diverse set of fermentation pathways while the second model (iFermGuilds564) separates fermentation activities into functional guilds. Both models predicted an energetic advantage for production of octanoic acid (a desirable MCFA) as a major fermentation product. In simulating behavior of our bioreactor microbiome, iFermGuilds564 predicted that MCFA were largely produced by a sugar-consuming guild, while short-chain fatty acids were mainly produced by a guild consuming lactate. Other guilds were predicted to ferment sugars and complex carbohydrates to lactate, acetate, H_2, and CO_2. Acetate, H_2, and CO_2 are all predicted to be terminal products rather than MCFA intermediates. Using iFermGuilds564, we also predicted that a community of MCFA-producing sugar and acetate consumers and homoacetogens could improve production of octanoic acid. In conjunction with past metagenomic and metatranscriptomic data, these results provide an improved understanding of MCFA production. The first generation metabolic models for MCFA production (https://github.com/mscarbor/Mixed-Culture-Fermentation-Models) represent a step towards predictive models for designing microbiomes to produce MCFAs.

An Investigation into Ratiometric Control for Synthetic Ecosystems

Authors:

Boo, A., Imperial College London

Ledesma-Amaro, R., Imperial College London

Stan, G. B., Imperial College London

In nature, cooperation between microbes allow microbial ecosystems to resist through fluctuating environmental conditions and produce complex substances necessary for their survival. Synthetic biologists want to harness the potential of such ecosystems for their own purposes and thus are investigating ways to build stable synthetic ecosystems. However their synthetic consortia often grow out of balance when some strains present growth advantages and outcompete the remaining strains in the consortia, leading to a loss of function of their systems. Here we design gene circuits in *E. coli* to achieve growth regulation that do not rely on the use of gene knockouts in order to conserve optimal growth properties and activate the controller only when the consortium becomes unstable. We focus on the development of an RNA-based controller capable of sensing cell density and regulating growth in the context of host-aware synthetic biology. This investigation is a first step in understanding how RNA could be used as a highly engineerable platform to control growth rates in stable synthetic ecosystems for applications in metabolic engineering.

Bacterial Modification of Bile Acids Alters Host Metabolism

International Conference
MICROBIOME
ENGINEERING

Authors:

Devlin, S., Harvard

Bile acids are steroidal metabolites that play crucial roles in host metabolism by acting both as detergents that aid in digestion and as ligands for host receptors, including nuclear hormone receptors and GPCRs. In the colon, bacteria chemically modify host-produced primary bile acids, generating a group of approximately 50 metabolites called secondary bile acids. While imbalances in this bile acid pool have been correlated with metabolic disorders, it is unclear which bile acids exert beneficial or harmful effects on host metabolism. My lab can now control the in vivo pool of bile acids in mice by inhibiting the keystone enzyme of bacterial bile acid metabolism, bile salt hydrolases (BSHs), using both genetic and chemical tools. We find that knocking out a selective BSH in monocolonized mice results in altered host metabolism, including reduced weight gain, reduced respiratory exchange ratios, and altered transcriptional profiles in the gut and liver. These results indicate that the deletion of a single bacterial gene can profoundly alter host physiology. In ongoing work, we have developed a small molecule inhibitor of BSHs, and we are using this compound as a chemical tool to probe the effects of bacterial bile acid modifications on host metabolism in fully colonized animals.

Ecosystem fabrication: New strategies for assessing microbial interactions in plant and soil microbiomes

Authors:

Andeer, P., Lawrence Berkeley National Laboratory

The ability to engineer microbiomes is desirable both to understand how microbial communities function and increasingly as a strategy to alter in situ environments. Analogous to how probiotics are being used to improve human health, microbial amendments are sought after to improve plant and soil health. However, the variability and complexity of these ecosystems coupled with our lack of understanding of microbial interactions limit scientific examination and engineering approaches.

Most of what we know about microorganisms, including their gene functions and growth dynamics have been derived from single-organism studies in benchtop experiments or from correlations based on observations made in complex systems. To bridge the gap between fully-constrained single organism laboratory experiments and high-dimensional ecosystem analyses, we have been developing EcoFABs (Ecosystem Fabrication; www.eco-fab.org; https://doi.org/10.3791/57170) which are small chambers constructed using common microfluidic procedures that can be used to study and engineer microbiomes living on and around the root surface. The devices are designed to be amenable to a number of analyses including microscopy of the root zones and soils, metabolomics analyses, and spatial sampling of the microbial communities. Over a dozen different plant varieties have been grown in EcoFABs, and a number of growth conditions are supported including hydroponics, solid substrate and live soils.

This focus on rhizosphere microbiomes is both based on their importance in agriculture and the fact that the root surface provides a reproducible and relevant habitat for laboratory experimentation. Recently, a multi-laboratory EcoFAB study growing the model plant Brachypodium distachyon demonstrated reproducible plant phenotypes and metabolomics including significant differences between treatments that were retained across laboratories (https://doi.org/10.1111/nph.15662). Current efforts are focused on the development of fully defined fabricated ecosystems that can be disseminated between laboratories. These include the EcoFAB devices, standardized microbiomes, protocols, equipment and software.

Depletion of Microbiome-Derived Molecules in the Host Using Clostridium Genetics

International Conference
MICROBIOME
ENGINEERING

Authors:

Guo, C. J., Weill Cornell Medicine

Fischbach, M., Stanford University

The rapid expansion of research using CRISPR-Cas requires a flexible software solution to guide precise genome editing beyond model organisms and simple applications. Here we present CASPER (CRISPR Associated Software for Pathway Engineering and Research), an integrated software platform that provides end-users the ability to design CRISPR tools to manipulate non-traditional biological systems beyond model organisms with a wide-variety of newly discovered and engineered Cas endonucleases. CASPER contains algorithms for on- and off-target analysis and multitargeting that enable target design of user-provided genomes and populations/metagenomes. Using CASPER, we were able to rapidly construct a library of 125 targets for a *S. aureus* infection in a skin microbiome, limiting predicted off-targets across the microbial community and cross-compatibility with both the endogenous CRISPR-Cas system of *S. aureus* and the ubiquitous *S. pyogenes* Cas machinery. Tight integration with NCBI and KEGG databases expands the bioinformatic capabilities of CASPER, including the ability to rapidly generate guide-RNA libraries from database provided annotations.

Development of an Engineered Therapeutic E. coli Nissle for the Treatment of Phenylketonuria

International Conference
MICROBIOME
ENGINEERING

Authors:

Charbonneau, M., Synlogic Inc

Phenylketonuria is a monogenic disease characterized by the inability to metabolize phenylalanine (Phe), which can result in neurotoxicity. Standard treatment for PKU comprises severe dietary restriction of protein to control plasma Phe levels. There is an unmet medical need for an orally administered therapeutic that can metabolize dietary Phe in the gastrointestinal tract and reduce plasma Phe. To address this problem, we engineered a strain of Escherichia coli Nissle 1917, termed SYNB1618, to produce two distinct Phe metabolizing enzymes: phenylalanine ammonia lyase (PAL) and L-amino acid deaminase (LAAD). We also developed an in vitro simulation (IVS) platform that recapitulates physiological parameters of the human upper gastrointestinal tract. Under IVS conditions, SYNB1618 consumed Phe over a period of several hours, consistent with human upper gastrointestinal transit. The product of PAL, trans-cinnamic acid (TCA) is quantitatively converted to hippuric acid (HA) in vivo and excreted in the urine, enabling the use of HA as a predictive biomarker of SYNB1618 activity. In non-human primates, administration of SYNB1618 resulted in dose-dependent increases in serum TCA and urinary HA. Next, we constructed a mechanistic in silico model considering human upper gastrointestinal physiology as well as IVS characterization of SYNB1618. Monte Carlo simulations were used to model SYNB1618 dose escalation in humans, corresponding to a dose-dependent increase in predicted Phe consumption. A Phase 1 clinical study revealed that orally-administered SYNB1618 was safe and generally well tolerated by healthy volunteers and PKU patients. In addition, dose-dependent recovery of TCA and HA was observed with SYNB1618 administration in plasma and urine, respectively. Notably, urinary HA recovery in humans closely mirrored predictions made by our in silico model. These studies indicate that SYNB1618 can impact Phe metabolism in a dose-dependent manner in preclinical models and humans, supporting the potential for the treatment of metabolic disorders with engineered bacteria.

Engineering Bacterial Therapy in Multicellular Coculture Model

Authors:

Harimoto, T., Columbia University

Singer, Z., Columbia University

Velazquez, O., Columbia University

Zhang, J., Columbia University

Castro, S., Columbia University

Hinchliffe, T., Columbia University

Mather, W., University of California, San Diego

Danino, T., Columbia University

Engineering bacteria to locally supplant their natural niches and modify disease state is emerging area of synthetic biology. However, bacterial behavior in complex physiological environment remains largely uncharacterized due to lack of spatiotemporal information attained from animal models. We recently developed a coculture platform that stably grow various bacteria within the necrotic core of multicellular spheroids over days, recapitulating selective bacterial growth in solid tumor *in vivo*. Using time-lapse microscopy, we quantitatively measure dynamics of bacterial tumor colonization over time and space. We observed differential kinetics and spatial distribution of bacterial invasion in tumor mass among the bacterial species. Utilizing this system, we screened tumor-targeting bacteria that deliver a library of anticancer molecules via synthetic gene circuits and identified candidates exhibiting significant tumor reduction. We hope this approach will be useful as a starting point towards implementing engineered bacteria for *in vivo* therapeutic applications.

Engineering Commensal Clostridia for Therapeutic Microbiome Interventions

International Conference
MICROBIOME
ENGINEERING

Authors:

Heeg, D., CHAIN Biotechnology Ltd, Koelschbach, J., CHAIN Biotechnology Ltd, Harding, E., CHAIN Biotechnology Ltd, Bradley, B., CHAIN Biotechnology Ltd, Green, E., CHAIN Biotechnology Ltd

Commensal gut bacteria contribute a large set of genes and their products to their host, influencing a variety of functions such as metabolism and defence from pathogens. Recent research has associated disruption of the homeostatic relationship between the human host and the gut microbiota with a variety of morbidities, ranging from inflammatory bowel diseases to metabolic and even neurological disorders. There is, therefore, enormous interest to explore the potential use of host microbes as therapeutic interventions for many human diseases. Probiotic bacteria have been used for many years to alleviate conditions linked to gastrointestinal (GI) distress and pathology, and more recently faecal microbiota transplant for treatment of antibiotic-associated diarrhoea and recurrent *C. difficile* associated disease has further highlighted the potential for gut microbiome-based interventions. While such approaches using wildtype strains show clinical efficacy relating to the natural properties of these strains, the possibility to genetically engineer these gut bacteria to optimise specific functional traits opens up an even greater opportunity. CHAIN Biotechnology Ltd is genetically engineering a member of Clostridia, which due to its beneficial properties, including production of the SCFA butyrate, and its sporulating nature has long been used as an effective probiotic to treat GI dysfunction. CHAIN's platform strain can be engineered to produce a variety of therapeutic products, including metabolites, anti-inflammatory and anti-bacterial molecules. Being administered in spore form, the engineered strains have advantages over other therapeutics relating to ease of delivery, manufacture, storage and shelf-life. Here, we report on progress made with the development of the platform strain and its derivatives engineered to produce therapeutics for treatment of inflammatory bowel disease, antibacterial treatment, and in mucosal vaccination strategies for viral and bacterial infections.

Engineering Living Medicines for Chronic Diseases

International Conference Microbiome Engineering

Authors:

Whitaker, W., Novome Biotechnologies

Shepherd, E., Novome Biotechnologies

DeLoache, W. C., UC Berkeley

Russ, Z. N., UC Berkeley

As the molecular mechanisms underlying the associations between the microbiome and diseases are discovered, delivering defined activities to gut will become an increasingly important therapeutic modality. A key challenge for microbiome therapeutics is delivering predictable, high doses of therapeutics strains despite widely varying patient microbiomes. We have developed a suite of synthetic biology tools that enable the programing of therapeutic activities into the most abundant genus of gut bacteria, Bacteroides. Our therapeutic Bacteroides strains are engineered to consume a rare carbohydrate, porphyran, so that even within a complex community we can exclusively feed them with a prebiotic. By creating this exclusive niche, we can colonize diverse, and even resistant, microbiotas at a predictable level that is tunable with porphyran dosing. To avoid the uncontrolled release of engineered strains we have further modified the strains to rely on porphyran for function of essential genes, so that strains are only viable in the subject during porphyran dosing. We are applying our platform for robustly and reversibly colonizing the gut with engineered bacteria to treating enteric hyperoxaluria, a chronic disease where over absorption of oxalate in the colon can result in recurrent kidney stones. Inspired by natural oxalate degrading bacteria often found in the gut at low levels, we have engineered rapid oxalate degradation into our robustly colonizing Bacteroides strain. When introduced in rat and pig models of enteric hyperoxaluria, our strain successfully reduces urine oxalate. We are currently working to bring this therapeutic into the clinic and are investigating treatment of other microbiome associated disease.

Synthetic Approaches to Engineering Phage Transduction

Authors:

Berry, J. D., Caribou Biosciences, Inc., Oromí-Bosch, A., Caribou Biosciences, Inc., Beljouw, S. V., Caribou Biosciences, Inc., Kotula, J. W., Caribou Biosciences, Inc., Owen, A., Caribou Biosciences, Huang, J., Caribou Biosciences, Inc., Shaw, M., Caribou Biosciences, Inc., Smith, S., Caribou Biosciences, Inc., Zhou, X., Caribou Biosciences, Kanner, S., Caribou Biosciences, Inc.

Commensal microbes from the human gastrointestinal (GI) tract have been shown to directly impact human health. We can magnify their beneficial effects by editing the genomes of these organisms in order to control when and where specific metabolic pathways are activated, thereby creating controllable, living therapeutics. Current tools do not enable effective genetic manipulation of most gut microbes. DNA delivery into uncultivated bacteria remains a significant initial barrier to genetic manipulation of commensal bacteria. Bacteriophages, bacterial viruses that have evolved to deliver their DNA into bacteria, represent an exciting tool to overcome this obstacle. At Caribou Biosciences, Inc., we are developing a platform for phage manipulation to alter surface receptor specificity and enable plasmid delivery to specific bacterial targets.

Data is presented describing our efforts utilizing bioinformatics, DNA synthesis, NGS assays, and Cas9-based phage manipulation to modify DNA packaging and tropism interactions. The primary focus is on engineering phages *in vitro* through the use of cell-free systems (CFS). We engineer a phage genome with Cas9 RNPs to modify host range via CFS-based genetic complementation. We demonstrate packaging of Cas9-based tools in a CFS and delivery of those tools to a host microbe. These achievements, along with other ongoing work, will pave the way to significantly decrease the time required to successfully engineer bacteria isolated from the human GI tract.

Towards Engineered Probiotics to Deliver Narrow-Spectrum Antimicrobial Peptides Against Drug-Resistant Enteric Bacteria

International Conference
MICROBIOME
ENGINEERING

Authors:

Mortzfeld, B. M., University of Massachusetts Medical School

Palmer, J. D., University of Oxford

Dupre, H. L., University of Massachusetts Dartmouth

Bucci, V., University of Massachusetts Medical School

Worldwide medical complications related to multi-drug resistant Enterobacteriaceae are a major issue in modern healthcare.

Working towards the development of antibiotic alternatives, we utilize the ability to heterologously produce the antimicrobial peptides microcin H47 (MccH47) and I47 (MccI47) in order to inhibit growth of drug-resistant Enterobacteriaceae in static plate assays. Moreover, using a Maltose Binding Protein (MBP)-microcin fusion protein system we are able, for the first time, to purify H47 near homogeneity and analyze in liquid minimum inhibitory concentration (MIC) assays. Our data reveal different, but overlapping, spectrums for both microcins with high specificity of H47 for candidate members of *E. coli, Shigella spp.*, *Salmonella spp.* and *Proteus mirabilis,* including multi-drug resistant isolates. Previously uncharacterized I47 observes high antimicrobial activity against *Klebsiella pneumoniae*.

Engineering probiotic strains like *E. coli* Nissle 1917 with the ability to overproduce these antimicrobial compounds provides a novel strategy for the treatment of antibiotic resistant pathogenic infections.

An Agent-Based Modelling Framework to Investigate the Role of Chlamydia Pneumoniae in Late-Onset Alzheimer's

International Conference
MICROBIOME
ENGINEERING

Many microbiomes exhibit remarkable diversity, even within the same ecosystem. The mechanisms that enable such broad coexistence are critical to understanding the composition and function of microbiomes. Network structure (e.g. cross-feeding) or spatial structure (e.g. compartmentalization) are well-studied mechanisms that promote coexistence in microbial communities. Temporal structure, in which nutrients or stresses change over time in an ecosystem, is a third mechanism that can promote coexistence between competing microbes. An improved understanding of temporal structure could unlock new approaches to tune the diversity or composition of microbiomes *in-situ*. In prior work, we developed eVOLVER, a DIY automated culture platform that enables real-time monitoring and feedback control over culture parameters (e.g. culture density, temperature, media composition) across an array of independent cultures for long-term experiments (>100 hours). Recently, eVOLVER enabled us to vary the temporal structure in microbial co-cultures by independently tuning mean dilution rate (i.e. cumulative media consumption) and dilution frequency (i.e. from 'batch-Like' to 'chemostat-like'). Starting >100 continuous cultures from the same highly diverse soil microbiome sample, we observed large differences in the composition and diversity of communities grown in different dilution schemes. Notably, our results diverged in multiple ways from the classic diversity-disturbance relationship models that are used to relate environmental fluctuations to ecosystem diversity. In collaborative work, we found our observations are well-captured by a modified consumer-resource model. We believe these results reinforce the role of temporal structure in stabilizing microbial communities, and highlight the need to consider these effects in both experimental and modeling efforts to study and engineer microbiomes.

Anaerobic Microbial Consortia Omics to Design Agile and Expedient Bioreactors

Authors:

Perisin, M., CCDC ARL, Leary, D., Naval Research Laboratory, Benyamin, M., Oak Ridge Associated Universities, Schwalm, N. III, CCDC ARL, Renberg, R., GTS, Gerlach, E., CCDC ARL:, Mojadedi, W., ORAU, Liu, S., U.S. Army Research Laboratory, Sund, C., U.S. Army Research Laboratory, Katherine, A., CCDC ARL

To win in the deep operational environment, the Army will need to be more adaptive and expeditionary with less logistic demand. Readily available resources such as food waste and vegetation provide waste streams from which to generate energy and materials at the point of need. These substrates are poor candidates for chemical conversion; however, microbes such as the inhabitants of human and ruminant guts can ferment these waste streams to commodity/specialty chemicals. Industrial fermentations using these microbes generally involve monocultures with highly controlled conditions that are inflexible to changing waste streams and environmental conditions. We propose to overcome these challenges with reconfigurable microbial reactors that customize consortia composition to direct variable waste inputs to tunable products. We hypothesize that consortia will outperform monocultures for waste to chemical production in Army relevant conditions. To assess consortia vs monocultures for chemical production from lignocellulosic feedstocks, we experimentally combined ruminant gut fungi with *Clostridium acetobutylicum (Cac)*. The fungi degrade lignocellulose to release carbohydrates that *Cac* then converts to useful chemicals. This carbohydrate mixture contains substrates with varying redox states known to affect *Cac* metabolic outputs. We further characterized the transcriptome and proteome of *Cac* in the presence of these substrates in order to understand their effects on metabolic regulation. The data was incorporated into a genome-scale metabolic model to better predict metabolic outputs and understand how genetic manipulations could direct useful chemical production given variable substrate inputs. Thus, we provide a theoretical and experimental framework to design, test, and build functional microbial consortia for expeditionary force sustainment.

Bacterial Community Modeling Allows Metabolic Differentiation of Recurrent and Nonrecurrent Clostridioides Difficile Infections in the Human Gut

Authors:

Henson, M. A., University of Massachusetts Amherst

Abstract

The anaerobic bacterium *Clostridioides difficile* is an opportunistic human pathogen responsible for infections of the colon of the human gastrointestinal tract. Recurrence is a major challenge associated with *C. difficile* infection (CDI), as approximately 30% of patients develop a least one occurrence of reinfection. A recent study compared the microbial communities of 275 stool samples collected from 93 patients with recurrent and nonrecurrent CDI. Bacterial taxa abundance data showed that patients who suffered from recurrent CDI harbored less diverse communities that were less likely to recover following treatment compared to patients with nonrecurrent CDI.

We hypothesized that integrating abundance data within community models could provide insights into the metabolic signatures of recurrent CDI. We assigned 16S rRNA-encoding gene sequence reads for the top 100 operational taxonomic units to 48 taxa (genera and families) and normalized the reads to produced taxa abundances for each sample. Direct analysis of these normalized abundances showed significant differences in five taxa, including elevated *Peptostreptococcaceae* (contains *C. difficile*) in recurrent samples and elevated *Bacteroides* in nonrecurrent samples. To better understand the functional impact of these compositional differences, the mgPipe metagenomics modeling tool was used to create and simulate 275 sample-specific community models with taxa abundances imposed as constraints. Model predictions showed significant differences in the ability of communities from untreated recurrent and nonrecurrent patients to produce 13 metabolites. Five degradation products from Stickland metabolism of aromatic amino acids were elevated in recurrent patient samples, including the potent bacteriostat para-cresol known to be secreted by *C. difficile* and toxic to gram-negative commensals such as *Bacteroides*. Conversely, the intermediate chorismite from the shikimate pathway for aromatic amino acid synthesis was elevated in nonrecurrent samples. These *in silico* predictions support recent studies suggesting aromatic amino acid metabolism as a key community function for understanding and potentially treating CDI.

Casper (CRISPR-Associated Software for Pathway Engineering and Research): A Versatile Tool for Harnessing CRISPR-Cas in Microbiome Engineering

Authors:

Mendoza, B., University of Tennessee Knoxville

Trinh, C. T., University of Tennessee

The rapid expansion of research using CRISPR-Cas requires a flexible software solution to guide precise genome editing beyond model organisms and simple applications. Here we present CASPER (CRISPR Associated Software for Pathway Engineering and Research), an integrated software platform that provides end-users the ability to design CRISPR tools to manipulate non-traditional biological systems beyond model organisms with a wide-variety of newly discovered and engineered Cas endonucleases. CASPER contains algorithms for on- and off-target analysis and multitargeting that enable target design of user-provided genomes and populations/metagenomes. Using CASPER, we were able to rapidly construct a library of 125 targets for a *S. aureus* infection in a skin microbiome, limiting predicted off-targets across the microbial community and cross-compatibility with both the endogenous CRISPR-Cas system of *S. aureus* and the ubiquitous *S. pyogenes* Cas machinery. Tight integration with NCBI and KEGG databases expands the bioinformatic capabilities of CASPER, including the ability to rapidly generate guide-RNA libraries from database provided annotations.

Detecting Changes in The C. Elegans digestive Tract Using an Engineered Bacterial Biosensor

Authors:

Rutter, J., University College London

Ozdemir, T., University College London

Barnes, C. P., University College London

Quintaneiro, L., Birbeck

Rosa, L., UCL

Cabreiro, F., LMS Imperial College London

Thomas, G., UCL

Galimov, E., LMS Imperial College London

Caenorhabditis elegans has become an important model organism within biology. The transparent gut, rapid growing time and ability to create a defined gut microbiota make it an ideal candidate organism for understanding and engineering the host microbiota. Here we present the development of an experimental model which can be used to help characterise whole-cell bacterial biosensors *in vivo*. A dual-plasmid biosensor system responding to IPTG was developed and fully characterised *in vitro*. Subsequently, we developed an automated image analysis pipeline and showed the sensor was capable of detecting and reporting on changes in the intestinal environment of *C. elegans*, after introducing exogenous biosensor inducer into the environment. The protocols presented here may be used for aiding the rational design of engineered bacterial circuits, primarily for diagnostic applications. In addition, the model system may serve to reduce the use of current animal models and aid in the exploration of complex questions within general nematode and host-microbe biology.

Development of Designer Probiotics for Targeted Delivery of Immunomodulatory Payloads

Authors:

Lynch, J., Massachusetts General Hospital

Drug delivery platforms that target the deposition of therapeutics to sites of disease are needed in order to increase efficacy and decrease off-target effects of current interventions. To address these issues, we have developed a probiotic *E. coli* capable of secreting therapeutic proteins into its surroundings. Our chosen chassis, *E. coli* Nissle 1917 (EcN), a probiotic with a well-established safety record in humans, can colonize the intestines as well as solid tumors. To enable EcN to secrete therapeutic payloads, we equipped it with a modified type III secretion system (T3SS), a protein delivery system common to many pathogenic Gram-negative bacteria. Using recombineering-based approaches, we transferred the operons that encode the core *Shigella* T3SS onto the chromosome of EcN. Next, we developed variants of this strain, referred to as T3EcN, that secrete functional nanobodies, variable domains of heavy chain only antibodies (aka VHH) that were isolated from immunized alpacas and identified to sequester and neutralize the target protein of interest (e.g. pro-inflammatory cytokines or immune checkpoint molecules). This was done by appending the N-termini of the nanobody with a type III secretion sequence. To enable T3EcN to stably colonize and constitutively secrete nanobodies *in vivo*, the nanobody-coding genes were placed under the control of constitutive synthetic promoters and introduced onto a plasmid that can be maintained in the absence of antibiotic selection. Notably, we observe that nanobody-secreting T3EcN variants stably colonize and maintain secretory activity within the mouse gut as well as tumors. Furthermore, using a bioluminescent reporter for T3SS activity, we have demonstrated that T3EcN's T3SS is active in the colon and small intestine. Ongoing work in preclinical murine models is focused on developing proof of concept for this novel targeted drug delivery platform.

Engineering a Microbiome Malathion Sensor Using Data-Driven Approaches

Authors:

Hasnain, A., UCSB

Yeung, E., Pacific Northwest National Laboratory

Malathion is a chemical compound commonly found in a wide variety of pesticides throughout the world. It is known to be toxic to birds and aquatic species, as well as slightly toxic to humans. In this work, we engineer a microbiome to act as bacterial sensors in various strains of bacteria to act as a Malathion detector. When Malathion is detected, the microbiome is engineered to fluoresce. Our methodology consists of first collecting time-series RNA sequencing measurements of a microbiome of Vibrio natrigens, Pseudomonas fluorescens, and Burkholderia Thailandensis as they are cultured in LB broth spiked with 0.615 mg/L of Malathion and also another batch cultured without any Malathion input. It is important to note that the growth curves of these strains growing in LB broth with and without Malathion were identical. From the gene expression measurements, we then identify the locations of the bacterial genomes which respond strongly to the Malathion chemical input using data-driven, Koopman operator approaches. Although the gene expression dynamics are highly nonlinear, our Koopman operator approach allows us to lift the dynamics to a higher-dimensional space where the dynamics are linear. In this new observable space, we make use of the observability gramian to identify the most observables modes of the system.

Functional Anabolic Network Analysis of Human-Associated Lactobacillus Strains

Authors:

Moutinho, T. J. Jr., University of Virginia

Papin, J., University of Virginia

Neubert, B. C., University of Virginia

Jenior, M., University of Virginia

Carey, M. A., University of Virginia

Medlock, G. L., University of Virginia

Kolling, G., University of Virginia

Members of the *Lactobacillus* genus are frequently utilized in the probiotic industry with many species conferring demonstrated health benefits; however, these effects are largely strain-dependent. We designed a method called PROTEAN (Probabilistic Reconstruction Of constituent Anabolic Networks) to computationally analyze the genomic annotations and predicted metabolic production capabilities of 144 strains across 16 species of *Lactobacillus* isolated from human intestinal, oral, and vaginal body sites. Using PROTEAN we conducted a genome-scale metabolic network comparison between strains, revealing that metabolic capabilities differ by isolation site. Notably, PROTEAN does not require a well-curated genome-scale metabolic network reconstruction to provide biological insights. We found that predicted metabolic capabilities of lactobacilli isolated from the vaginal microbiota cluster separately from intestinal and oral isolates, and we also uncovered an overlap in the predicted metabolic production capabilities of intestinal and oral isolates. Using machine learning, we determined the most informative metabolic products driving the difference between predicted metabolic capabilities of intestinal, oral, and vaginal isolates. Notably, intestinal and oral isolates were predicted to have a higher likelihood of producing D-alanine, D/L-serine, and L-proline, while the vaginal isolates were distinguished by a higher predicted likelihood of producing L-arginine, citrulline, and D/L-lactate. We found the distinguishing products to be consistent with published experimental literature. This study showcases a systematic technique, PROTEAN, for comparing the predicted functional metabolic output of microbes using genome-scale metabolic network analysis and computational modeling and provides unique insight into human-associated *Lactobacillus* biology.

High Order Bacterial Interactions Influence Activity of a Potential Bacteriotherapy Agent

Authors:

Cervantes, B., Massachusetts Institute of Technology

Bacterial replacement therapy has been proposed as a strategy to reduce carriage of antibiotic resistant (ABr) microbes. This type of bacteriotherapy works by providing a strain or a combination of strains capable of outcompeting the ABr microbe from its niche and subsequently occupying the same niche. Here we develop tools that facilitate the identification of potential bacteriotherapy strains and study the influence of environmental factors on their efficacy. We first describe a technique to rapidly study pairwise interactions between bacterial isolates that uses time lapse imaging of bacterial colonies in pairwise combinations. We use this technique to study isolates from healthy human stool against known antibiotic resistance carrying pathogens. We identify a potential bacteriotherapy strain against *Enterococcus faecium*, a known carrier of vancomycin resistance. The strain is lethal against clinical isolates of *E. faecium* and it is also hypothesized to occupy a similar metabolic niche. Then we study the influence of third party microbes on the efficacy of our potential bacteriotherapy strain. Interestingly, the presence of certain third party microbes can significantly change the efficacy of our potential bacteriotherapy. Our results begin to describe mechanisms that influence the robustness of our bacteriotherapy strain and shed light on prospective generalizable criteria for the selection of bacterial replacement strains.

High-Throughput Characterization of Host-Bacteriophage Interactions with an Eye Towards Emerging Pathogenic Threats

Authors:

Adler, B. A., UC Berkeley UCSF Graduate Program in Bioengineering

Zhong, C. Y., University of California, Berkeley

Deutschbauer, A. M., Lawrence Berkeley National Lab

Mutalik, V. K., Lawrence Berkeley National Lab

Arkin, A. P., University of California, Berkeley

Bacteriophage are diverse, critical players in the ecological dynamics and structure of microbial communities through the predation of their respective microbial hosts. Due to specific interactions with their microbial hosts, rational combinations of bacteriophage can be deployed to complex communities and eliminate targeted pathogens. Importantly, design principles are emerging for increased efficacy during such treatments rational phage cocktail design can reduce the risk of pathogen escape and increase population-level antibiotic sensitivity. However, leveraging these design principles into actionable treatments, products, and bottom-up engineering practices relies upon extensive characterization of how the bacteriophages interact with pathogen physiology. Recent work from our group and others demonstrated substantial increases in both the pace and resolution that we can infer bacteriophage-host interactions by use of barcoded, pooled genetic screens. Here, we extend random bar code transposon-site sequencing (RB-TnSeq) to Salmonella typhimurium to investigate thousands of genetic interactions with eleven of its bacteriophage, of which only four had substantial characterization prior to this work. We uncover receptor requirements for all eleven phages tested, including novel, unanticipated, complex receptor requirements that would not have been discovered without multiple, unbiased genetic screens. We additionally discover and characterize new host-side requirements for broad-host range phages, highlighting the importance of intracellular transcription and environmental conditions. This work comprises the largest, functional characterization of Salmonella bacteriophages, providing new avenues, considerations, and features for rational phage cocktail design against this emerging pathogen.

Investigating the Dynamics of Microbial Consortia in Spatially Structured Environments

Authors:

Gupta, S., University of Wisconsin - Madison

Gomez, M. M., UC Santa Cruz

Romero, P. A., University of Wisconsin-Madison

Venturelli, O. S., University of Wisconsin-Madison

Microbial communities are diverse networks of microorganisms that exhibit spatial organization on multiple length scales due to a complex interplay between microbial interactions and interactions with the environment. Spatial structuring of microbial communities has a major impact on system behaviors and responses to environmental perturbations. We designed a microfluidic platform, MISTiC (Mapping Interactions across Space and Time in Communities), to investigate how the spatial arrangement of microbes influences interaction networks and responses to temporally changing environmental stimuli using a combination of time-lapse microscopy and computational modeling. We demonstrate a mapping between spatial positioning and gene expression, and identify key biochemical parameters that determine the sensitivity of microbial interactions to spatial patterning. Our results show that distance establishes a low-pass filter that preserves signals that operate on longer timescales and enhances the fidelity of information transmission. We found that metabolic interactions are highly sensitive to spatial arrangements and resource availability, and can also change over time. In sum, we developed MISTiC to investigate how spatial positioning, temporal perturbations, and environmental factors shape microbial community behaviors. Our findings suggest generalizable principles that could be exploited to program the spatiotemporal behaviors of microbial consortia.

Microsplit: Single-Cell RNA-Sequencing Technology for Microbes

International Conference
MICROBIOME
ENGINEERING

Authors:

Kuchina, A., University of Washington, Seelig, G., University of Washington, Brettner, L., University of Washington Medical Center, DePaolo, W., University of Washington Medical Center

I present μSPLiT (microbial split-pool ligation transcriptomics), a high-throughput single-cell RNA-sequencing (scRNA-seq) method tailored specifically for bacterial communities. A technology based on combinatorial split-pool transcript indexing, μSPLiT easily scales to process hundreds of thousands of cells in a single experiment without need for specialized equipment. We show that μSPLiT can be used simultaneously on gram-positive and gram-negative bacteria and provides single-cell transcriptomic data that resolves heterogeneous functional states within bacterial communities.

ScRNA-seq methods have become a mainstream tool for profiling cell types and states in eukaryotes. However, because of technical challenges, these methods so far have not been compatible with bacteria. μSPLiT is uniquely suited to overcome these challenges since it does not require isolation of individual microbes in droplets or wells and allows sample multiplexing for parallel screening of experimental conditions. Scaling exponentially with the number of barcoding rounds, μSPLiT enables a massive increase in the number of cells that can be sequenced, which makes it uniquely suitable for profiling large and diverse microbial communities.

We validated μSPLiT to detect stress responses in a mixed-species consortium of gram-positive and gram-negative model organisms *E. coli and B. subtilis*, and profiled transcriptional changes in *B. subtilis* cultures transitioning from logarithmic to stationary phase of growth. We showed that μSPLiT can be used to resolve heterogeneous subpopulations of cells activating specific transcriptional programs in different conditions, providing a first to date unbiased single-cell transcriptomic overview of metabolic and developmental shifts in gene expression. Next, μSPLiT will be applied to resolve heterogeneous functional cell states and cell-to-cell variation in the complex multi-species natural communities such as the microbiota.

Multi-Species Co-Culture Platform to Enable Spatial Analysis of Microbial Communication

Authors:

Vaiana, C. A., Massachusetts Institute of Technology, Kim, H., Massachusetts Institute of Technology, Buie, C. R., Massachusetts Institute of Technology, Mayali, X., Lawrence Livermore National Laboratory, Kimbrel, J. A., Lawrence Livermore National Laboratory

Microbes naturally interact with neighboring organisms and environments in a "social network" of shared molecules. The collective consortia of organisms can often perform tasks more complex than isolated monocultures. Understanding community interaction will enable synthetic biologists to engineer designer communities that solve problems in the natural world. Typical co-culture studies rely on agar plate diffusion assays, or diffusion between two liquid chambers separated by a porous membranes. Herein, we demonstrate a multi-well porous hydrogel platform that enables bacterial multispecies co-culture. The hydrogel physically isolates individual cultures while allowing media and small molecule exchange. Recovery of liquid culture wells after growth can be fed into existing bio-analysis pipelines such as cytometry, mass spectroscopy, and sequencing to obtain spatially resolved information. Herein, we demonstrate the detection of biologically relevant molecule gradients through the device, including pH, small molecule inducers, and antibiotic gradients originating from a single seeded well. We also demonstrate the long-term co-culture of bacterial and algal cells within the device, and assess population changes in bacterial cultures with and without algae present.
These data suggest the use of a nanoporous growth chamber to study community-level interactions between microbes and their neighbors.

On the Adaptation and Evolution of Sulfonamide-Subsisting Bacteria

International Conference MICROBIOME ENGINEERING

Authors:

Deng, Y., The University of Hong Kong, Zhang, T., The University of Hong Kong

The catabolic genes conferring antibiotics assimilation in bacteria, dubbed as subsistome, can provide a source of potential resistance determinants. Additionally, the distribution of such machinery may detoxicate the micro-environment inhabited by these bacteria, benefiting susceptible pathogens sharing that surroundings.

Using founding microbial communities collected from different municipal sewage, we examined the adaptation and evolution of sulfadiazine-subsisting bacteria occurring across both microbial community level and gene level in laboratory microcosms. By metagenomic assembly of enrichment cultures collected from three alternative stable states (spanning 334 days), in which recurrent complete mineralization of sulfadiazine as sole carbon source were recorded, we uncovered > 9000 bacterial genomes that are recapitulated in 511 species. In line with 16S rRNA gene amplicon analysis, analysis based on genome bins showed that microbial composition of sulfadiazine-subsisting community was contingent on founding populations on pre-assembly states, not only because of compositional variance across founding populations, but also because of inconsistent selection of adaptive species from the initial bacterial pool. Selected populations possess both plastic and evolved adaptation strategies.

Despite evidences based on short-read metagenomic sequencing indicated that genomic capacity for sulfadiazine subsistence was discovered in phylogenetic diverse genome bins, by using long-read sequencing, we validated that the canonic sulfadiazine metabolic gene clusters have assembled primarily vertically in sulfadiazine-subsisting communities, and it sweep to be fixed in *Micrococcaceae* lineage. We further isolated individual strains of *Micrococcaceae* across multiple communities and illustrated two representative configurations of sulfadiazine metabolic gene clusters, which can reside next to each other on the same chromosome or locate separately either on chromosome or extra-chromosomal mobile genetic elements, documenting the adaptive radiation in the specialist sulfadiazine-subsisting bacterial lineage.

Optimisation of a Whole Cell Bacterial Biosensor for Propionate

Authors:

Dekker, L., Imperial College London

Ozdemir, T., University College London

Barnes, C. P., University College London

Propionate is a short chain fatty acid produced by microbial fermentation of dietary fibre in the gastrointestinal tract and plays a role in several metabolic and inflammatory conditions including inflammatory bowel disease. In *Escherichia coli*, *prpR* encodes the positive regulator of the *prpBCDE* promoter (P*prpB*) which is induced in the presence of propionate and is responsible for expression of the propionate catabolic genes (*prpBCDE*). PrpR specifically senses 2-methylcitrate (2-MC), an intermediate of the methylcitrate cycle used by this bacterium to convert propionate to pyruvate. The CIDAR MoClo assembly method was used to construct a whole cell propionate biosensor based on the pPro expression vectors that contain a gene encoding green fluorescent protein (GFP) under the control of the P*prpB* promoter and the *prpR* transcriptional gene, which is activated by 2-MC signalling the presence of propionate in the environment. The biosensor can only function in the presence of the prp*BCDE* operon on the chromosome of the host strain as it requires the production of enzymes needed to breakdown propionate to 2-MC. Our propionate biosensor exhibited a concentration dependent response to propionate. The biosensor was optimised by changing the strength of the promoter controlling the expression level of *prpR* followed by a directed evolution approach to simultaneously select for increased sensitivity, lower background expression and high dynamic range. This biosensor could play a valuable role in further research of the influence of propionate on the gut microbiome and the diagnosis of gut metabolic and inflammatory conditions.

Resolving Signatures of Engineering in S. Cerevisiae and Nonconventional Yeasts with Whole Genome Sequencing

International Conference
MICROBIOME
ENGINEERING

Authors:

Collins, J., Worcester Polytechnic Institute

Young, E., Worcester Polytechnic Institute

Engineered probiotics have the potential to diagnose and treat many diseases linked to the gastrointestinal (GI) tract; however, their therapeutic potential is fundamentally constrained by their ability to colonize desired GI niches and specifically respond to endogenous and extrinsic stimuli. As such, a method to non-invasively control the persistence and localization of engineered probiotics can greatly improve therapeutic outcomes for GI linked disorders, in addition to being an important basic-research tool for studying gut microbial communities. Magnetic fields are uniquely suited to address this challenge because of their ability to freely penetrate tissue, however the rapid decay of fields from a magnetic source limits their broader utility for targeting cells within the GI-tract. Here we demonstrate a new approach that uses micromagnets, in concert with an external magnetic-field, in order to enhance the retention of orally administered probiotics in the murine GI tract. We show that this approach enhances the colonization of probiotic *E. coli* Nissle in mice that were not treated with antibiotics. Our cellular localization assisted by magnetic particles (CLAMP) strategy can potentially be generalized to any cell (bacterial, mammalian) or drug-containing nanoparticle of interest, and can also be combined with existing non-invasive imaging modalities, thereby facilitating remote control and sensing of engineered cells at sites of interest *in vivo*.

Transcriptome-Guided Parsimonious Flux Analysis Improves Predictions with Metabolic Networks in Complex Environments

Authors:

Jenior, M., University of Virginia

Moutinho, T. J. Jr., University of Virginia

Papin, J., University of Virginia

The metabolic responses of bacteria to dynamic extracellular conditions drives not only the behavior of single species, but also entire communities of microbes. Over the last decade, genome-scale metabolic network reconstructions have assisted in our appreciation of important metabolic determinants of bacterial physiology. These network models have been a powerful force in understanding the metabolic capacity that species may utilize in order to succeed in an environment. Increasingly, an understanding of context-specific metabolism is critical for elucidating metabolic drivers of larger phenotypes and disease. However, previous approaches to use these platforms in concert with omics data to better characterize experimental systems have met challenges due to assumptions necessary by the various integration platforms or due to large input data requirements. With these challenges in mind, we developed RIPTiDe (Reaction Inclusion by Parsimony and Transcript Distribution) which uses both parsimony of overall flux and transcriptomic abundances to identify the most cost-effective usage of metabolism that also reflects the cell's investments into transcription. Additionally, in biological samples where it is difficult to quantify specific growth conditions, it becomes critical to develop methods that require lower amounts of user intervention in order to generate accurate metabolic predictions. Utilizing a metabolic network reconstruction for the model organism *Escherichia coli* str. K-12 substr. MG1655, we found that RIPTiDe correctly identifies context-specific metabolic pathway activity without supervision or knowledge of specific media conditions. We also assessed the application of RIPTiDe to *in vivo* metatranscriptomic data where *E. coli* was present at high abundances, and found that our approach also correctly predicts metabolic behaviors of host-associated bacteria. In the setting of human health, understanding metabolic changes within bacteria in environments where growth substrate availability is difficult to quantify can have large downstream impacts in our ability to elucidate molecular drivers of disease-associated dysbiosis across the microbiota.

Tuning Microbial Diversity Using Temporal Structure in High-Throughput Continuous Culture

International Conference
MICROBIOME
ENGINEERING

Authors:

Mancuso, C. P., Boston University

Khalil, A. S., Harvard University

Gore, J., Massachusetts Institute of Technology

Abreu, C., MIT

Lee, H., MIT

Many microbiomes exhibit remarkable diversity, even within the same ecosystem. The mechanisms that enable such broad coexistence are critical to understanding the composition and function of microbiomes. Network structure (e.g. cross-feeding) or spatial structure (e.g. compartmentalization) are well-studied mechanisms that promote coexistence in microbial communities. Temporal structure, in which nutrients or stresses change over time in an ecosystem, is a third mechanism that can promote coexistence between competing microbes. An improved understanding of temporal structure could unlock new approaches to tune the diversity or composition of microbiomes *in-situ*. In prior work, we developed eVOLVER, a DIY automated culture platform that enables real-time monitoring and feedback control over culture parameters (e.g. culture density, temperature, media composition) across an array of independent cultures for long-term experiments (>100 hours). Recently, eVOLVER enabled us to vary the temporal structure in microbial co-cultures by independently tuning mean dilution rate (i.e. cumulative media consumption) and dilution frequency (i.e. from 'batch-Like' to 'chemostat-like'). Starting >100 continuous cultures from the same highly diverse soil microbiome sample, we observed large differences in the composition and diversity of communities grown in different dilution schemes. Notably, our results diverged in multiple ways from the classic diversity-disturbance relationship models that are used to relate environmental fluctuations to ecosystem diversity. In collaborative work, we found our observations are well-captured by a modified consumer-resource model. We believe these results reinforce the role of temporal structure in stabilizing microbial communities, and highlight the need to consider these effects in both experimental and modeling efforts to study and engineer microbiomes.

Understanding the Dynamics of a Methanotroph-Cyanobacterium Coculture through Kinetic Modeling and Experimental Verification

Authors:

Wang, J., Auburn University, Badr, K., Auburn University, Hilliard, M., Auburn University, Roberts, N., Auburn University, He, Q. P., Auburn University, Kalyuzhnaya, M. G., San Diego State University, Beliaev, A. S., Pacific Northwest National Laboratory

Industrial, municipal, and agricultural waste streams contain stranded organic carbon, which can be converted into biogas through anaerobic digestion. It has been demonstrated that biogas has immense potential as a renewable feedstock for producing high-density fuels and commodity chemicals. However, the utilization of biogas presents a significant challenge due to its low pressure, high proportion of CO_2 and presence of contaminants such as H_2S, ammonia, and volatile organic carbon compounds. To tap into this immense potential, effective biotechnologies that co-utilize both CO_2 and CH_4 are needed. Using the basic metabolic coupling principles utilized by many natural consortia, we have demonstrated that photoautotroph- methanotroph co-cultures offers a flexible and highly promising platform for biological CO_2/CH_4 co-utilization.

From an engineering perspective, coupling methanotrophic metabolism to photosynthesis offers three major advantages for biological biogas conversion. First, exchange of *in situ* produced O_2 and CO_2 dramatically reduces mass transfer resistance of the two gas substrates, which can dramatically increases the growth of both strains; Second, *in situ* O_2 consumption removes inhibition on photoautotroph and eliminates risk of explosion; Third, interdependent yet compartmentalized configuration of the coculture offers flexibility and more options for metabolic engineering.

However, development of multi-organism platforms for commercial biogas conversion present significant challenges which center around our ability to control function and composition of species in the coculture. An essential tool for the optimization, design and analysis of the coculture based biogas conversion is the development and validation of kinetics models that can accurately describe and predict the co-culture growth under different conditions. In this work, using *Methylomicrobium buryatense - Arthrosipira platensis* as the model coculture system, we present the very first effort to quantitatively model the growth dynamics of the coculture. The validated kinetic model can accurate predict the coculture dynamics under different growth conditions.

AUTHOR INDEX

AUTHOR INDEX